AF575498

BACKYARD BIRDS
Blackbirds
by Rachael Barnes
BLASTOFF! READERS
1
BELLWETHER MEDIA • MINNEAPOLIS, MN

Blastoff! Readers are carefully developed by literacy experts to build reading stamina and move students toward fluency by combining standards-based content with developmentally appropriate text.

Level 1 provides the most support through repetition of high-frequency words, light text, predictable sentence patterns, and strong visual support.

Level 2 offers early readers a bit more challenge through varied sentences, increased text load, and text-supportive special features.

Level 3 advances early-fluent readers toward fluency through increased text load, less reliance on photos, advancing concepts, longer sentences, and more complex special features.

★ **Blastoff! Universe**

Reading Level

Grade K

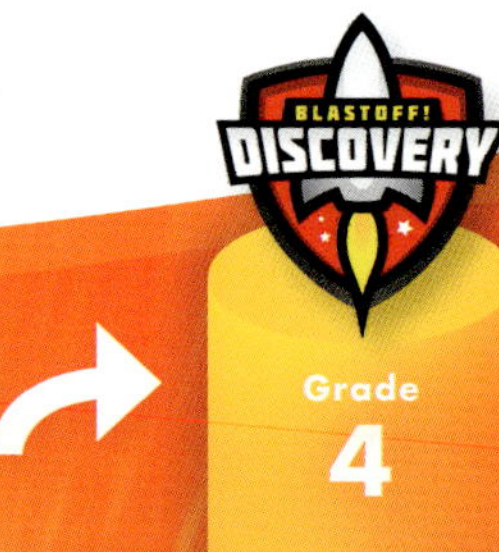

Grades 1–3

Grade 4

This edition first published in 2023 by Bellwether Media, Inc.

Library of Congress Cataloging-in-Publication Data

Names: Barnes, Rachael, author.
Title: Blackbirds / by Rachael Barnes.
Description: Minneapolis, MN : Bellwether Media, 2023. | Series: Backyard birds | Includes bibliographical references and index. | Audience: Ages 5-8 | Audience: Grades K-1 | Summary: "Developed by literacy experts for students in kindergarten through grade three, this book introduces blackbirds to young readers through leveled text and related photos"– Provided by publisher.
Identifiers: LCCN 2022002368 (print) | LCCN 2022002369 (ebook) | ISBN 9781644876893 (library binding) | ISBN 9781648347351 (ebook)
Subjects: LCSH: Blackbirds–Juvenile literature.
Classification: LCC QL696.P2475 B37 2023 (print) | LCC QL696.P2475 (ebook) | DDC 598.8/74–dc23/eng/20220125
LC record available at https://lccn.loc.gov/2022002368
LC ebook record available at https://lccn.loc.gov/2022002369

Editor: Rebecca Sabelko Designer: Laura Sowers

Printed in the United States of America, North Mankato, MN.

Table of Contents

What Are Blackbirds?

Blackbirds are large **songbirds**. They can have many colorful feathers!

All in the Family
Brewer's blackbird
red-winged blackbird
yellow-headed blackbird

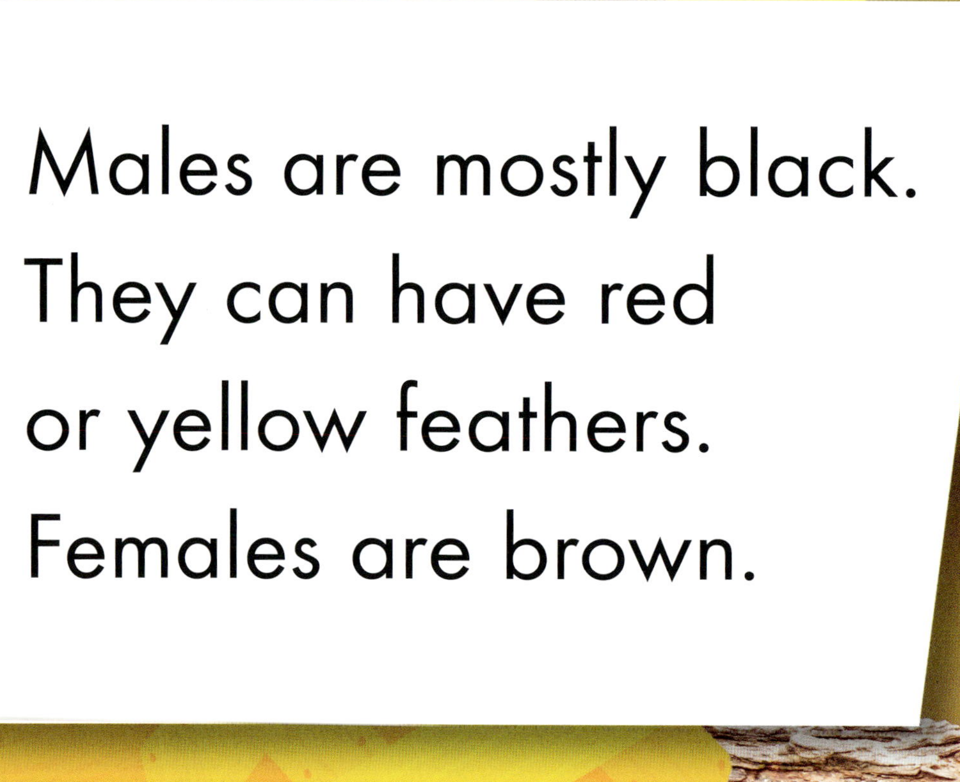

Males are mostly black. They can have red or yellow feathers. Females are brown.

female
males

Gathered in Groups

Blackbirds keep their **territories** safe. Some males watch for danger from **perches**.

perch

Female blackbirds build nests. The nests are often on tall plants in **wetlands**.

nest

Blackbirds look for food on the ground. They eat **insects** and seeds.

Blackbird Food
insects
seeds

Blackbirds eat in groups. Some fly over each other to reach seeds!

On the Move

Blackbirds make a lot of different sounds. Some calls are quick and loud!

Blackbird Call
conk-la-
ree!

Some blackbirds **migrate** south for winter. Others move west to warm shores.

migrating

Listen for their calls. Blackbirds are one of the first signs of spring!

Glossary

insects

small animals with six legs and hard outer bodies

songbirds

birds that make musical sounds

migrate

to travel with the seasons

territories

the land areas where animals live

perches

places where birds sit or rest high above the ground

wetlands

areas of land that are wet at least part of the year

To Learn More

AT THE LIBRARY

Devera, Czeena. *In My Backyard.* Ann Arbor, Mich.: Cherry Lake Publishing, 2019.

Mattern, Joanne. *Living in Swamps, Salt Marshes, and Other Wetlands.* New York, N.Y.: PowerKids Press, 2021.

Neuenfeldt, Elizabeth. *Orioles.* Minneapolis, Minn.: Bellwether Media, 2022.

ON THE WEB

FACTSURFER

Factsurfer.com gives you a safe, fun way to find more information.

1. Go to www.factsurfer.com.
2. Enter "blackbirds" into the search box and click 🔍.
3. Select your book cover to see a list of related content.

Index

The images in this book are reproduced through the courtesy of: Jeff Caverly, front cover (blackbird); Artazum, front cover (backyard); Agami Photo Agency, p. 3; Chon Kit Leong, pp. 4-5; Steve Byland, p. 5 (Brewer's blackbird); Glenn Price, pp. 5 (red-winged blackbird), 13 (seeds); Peter Gyure, p. 5 (yellow-headed blackbird); Thomas Torget, pp. 6-7; Ray Whittemore, p. 7 (females); Iv-olga, pp. 8-9; QualityHD, p. 9 (perch); Imagebroker/ Alamy, pp. 10-11; Amanda Cuercio, pp. 12-13; Costea Andrea M, p. 13 (insects); Bob Gibbons/ Alamy, pp. 14-15; JamesChen, pp. 16-17; kojihirano, pp. 18-19; Ami Parikh, pp. 20-21; yod 67, p. 22 (insects); Rabbitti, p. 22 (migrate); Lou Sisneros Photography, p. 22 (perches); Jordan Feeg, p. 22 (songbirds); Danita Delimont, p. 22 (territories); jaimie tuchman, p. 22 (wetlands); wildphoto3, p. 23.